CW01218752

IT'S WINE TIME

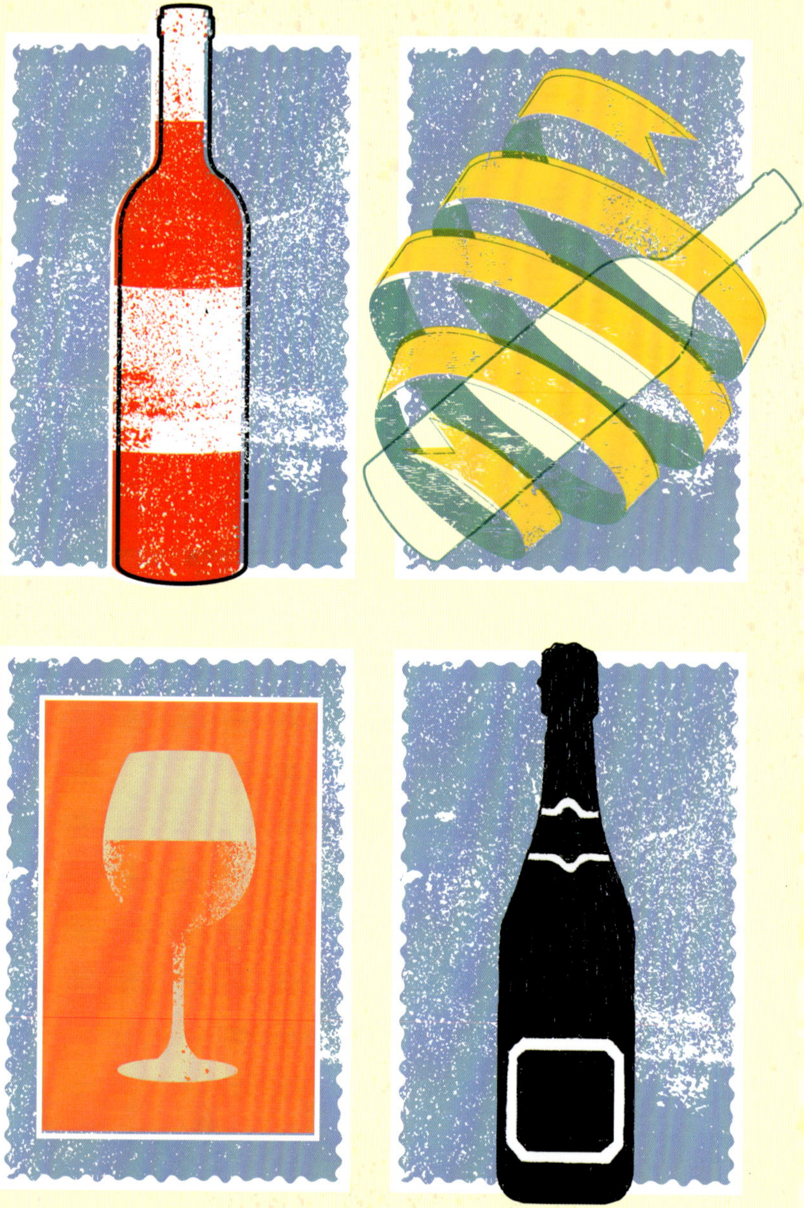

It's Wine Time

Everything you've always wanted to know but were too afraid to ask about red, white, rosé, and sparkling wine

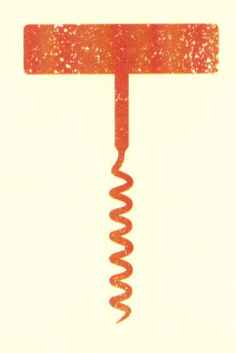

Chris Losh

DOG 'n' BONE

This edition published in 2020 by Dog 'n' Bone Books
An imprint of Ryland Peters & Small Ltd
20–21 Jockey's Fields
London WC1R 4BW
341 E 116th St
New York, NY 10029

www.rylandpeters.com

10 9 8 7 6 5 4 3 2 1

First published in 2005 by Ryland Peters & Small under the title *Pick The Right Wine Every Time*

Text © Chris Losh 2005, with the exception of Vegetarian dishes page 19:
Fiona Beckett © Ryland Peters & Small 2012
Design © Dog 'n' Bone Books 2020

The author's moral rights have been asserted. All rights reserved. No part of this publication may be reproduced, stored in a retrieval system, or transmitted in any form or by any means, electronic, mechanical, photocopying, or otherwise, without the prior permission of the publisher.

A CIP catalog record for this book is available from the Library of Congress and the British Library.

ISBN: 978 1 912983 23 0

Printed in China

Editor: Clare Double
Illustrations: Shutterstock/ZOO.BY; Shutterstock/ducu59us page 18; Shutterstock/helterskelter page 26; Shutterstock/lukeruk page 17; Shutterstock/Nataleana page 40; Shutterstock/ravennka pages 38 and 39; Shutterstock/Seita pages 49, 50, 51, 52, 53, and 54; Shutterstock/Sloth Astronaut pages 2, 24, and 39; Shutterstock/squarelogo pages 18, 19, 39, 44, and 45; Shutterstock/stockillustration page 12; Shutterstock/tsaplia page 32; Shutterstock/Vector1st page 39, and Shutterstock/wawritto pages 28 and 29

Commissioning editor: Kate Burkett
In-house designer: Emily Breen
Art director: Sally Powell
Head of Production: Patricia Harrington
Publishing manager: Penny Craig
Publisher: Cindy Richards

CONTENTS

Introduction 6

PICK RIGHT AT HOME 8
Wednesday night supper 10
Friday night binge-watching 12
before and after dinner 14
barbecue 16
Sunday lunch 18

PICK RIGHT ON SPECIAL OCCASIONS 20
dinner party 22
weddings and anniversaries 24
the holidays 26
party 28

PICK RIGHT OUT AND ABOUT 30
bars and pubs 32
date night 34
visiting the in-laws 36
picnics 38
gifts 40

PICK RIGHT EATING OUT 42
navigating the wine list 44
french 46
italian 48
chinese 50
japanese 51
thai 52
indian 53
mexican 54
middle eastern 55
seafood 56
fusion food 57

storing 58
serving 59

glossary of terms 60
index and acknowledgments 64

INTRODUCTION

Wine is the most versatile of drinks. It's your companion at home in front of the television, an ice-breaker at parties, and a gift for those you want to impress. It's something to be treated with reverence and with abandon, a drink for every budget and every occasion.

Wines, like people, have very different personalities, which is both their attraction and their curse. On one hand, the variety on offer makes wine more stimulating, varied, and fascinating than any other drink on the planet; on the other, with such a lot of choice it can be very easy to Get It Wrong.

The biggest difficulty nowadays is not so much getting a half-decent bottle (standards are far higher than they used to be), but in getting the right type of wine for the situation. Here, again, wines are like people.

You probably wouldn't want to receive investment advice from a party animal or be stuck in a bar with someone who talked about tax refunds all night, but put the accountant in his office and the reveler in his bar, and they come into their own. Well, so it is with wine. Take a perfectly good bottle and serve it at the wrong time or with the wrong food, and it might taste horrible, but get it in the right place at the right time and it will shine.

That's why this book is arranged not by wine style, but the way that most of us buy wine: by event. Whether it's a bottle for midweek glugging or Sunday lunch, a cookout or a blowout, the intention is to cut through the mystery and find wine styles that fit in with the way you live your life. There's nothing in here about magnificent Bordeaux châteaux or the history of winemaking in Australia; just practical information designed to help you get the most out of wine without spending a fortune. After all, wine might be stimulating and fascinating, but we don't need to make things any harder for ourselves than they already are.

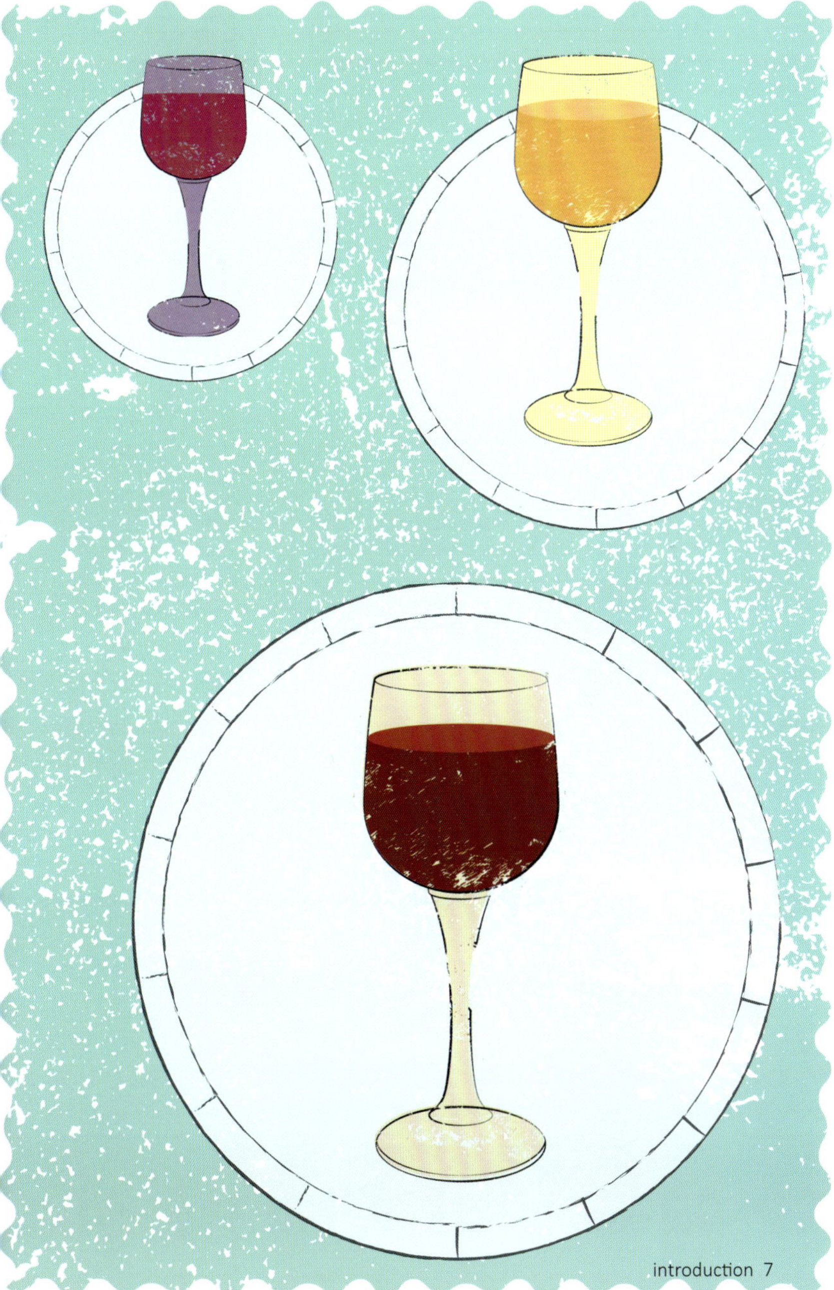

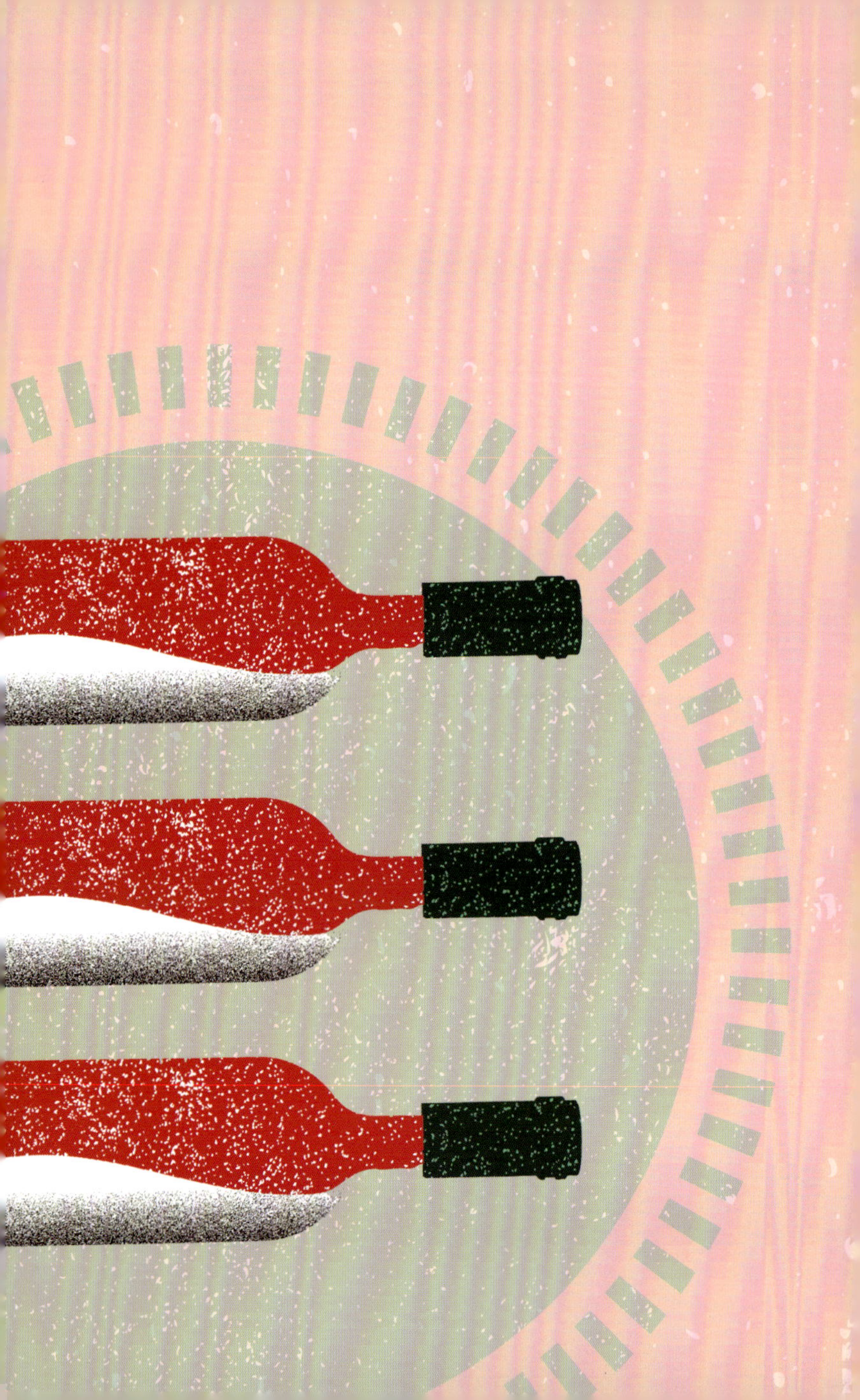

PICK RIGHT AT HOME

WEDNESDAY NIGHT SUPPER

Wine isn't just a special-occasion drink any more. With the possible exception of breakfast, it's a civilized accompaniment to more or less any meal you fancy, and the midweek couple of glasses are ever more popular.

For school nights, I'd suggest finding wine styles that will go with several dishes. That way, if you only drink a couple of glasses on one night, you've still got something left for the next day (see Keeping it Fresh, opposite).

Chardonnay, for instance, is a wonderfully versatile white. If it's not too heavily oaked, it'll go with fish, chicken, pork, and most pasta dishes. The same goes for other fleshy whites such as Pinot Blanc and Chenin Blanc. Sauvignon Blanc, on the other hand, might be a good match with fish, salad, and some oriental cuisines, but because it's lighter and more aromatic, it can struggle with white meats. The same goes for Pinot Grigio.

For reds, mid-weight wines like Pinot Noir and Rioja can work with everything from pasta to meaty fish, pork chops, and even chili. The same goes for young, unambitious Bordeaux. But open a big Cabernet Sauvignon or Shiraz, and, unless you drink it all in one sitting, it could be hanging around for a while. (I'm assuming here that you don't eat steak five nights a week!)

Because midweek food rarely takes more than half an hour to

KEEPING IT FRESH

★

Quite often I visit friends' houses to see half-drunk bottles with the cork stuck back in the top. Bad idea. If you only want to drink a couple of glasses from a bottle (understandable during the week), buy a vacuum set. This involves putting a rubber stopper in the top of the bottle and removing the air with a small pump. It's cheap, easy to use, lasts forever, and will keep your bottles fresh for a couple of days.

prepare, adjust your wine budget accordingly. Really cheap stuff might actually detract from your meal, but there's no point spending more than you have to. Chile, Argentina, and Portugal are good sources of decent, value-for-money reds; try Australia and South Africa for whites.

A good addition to the midweek kitchen is bag-in-box wine. It might not look pretty, but a three-liter box is usually cheaper than four bottles, you can have as much or as little as you want at a time, and the wine will stay fresh for several weeks.

FRIDAY NIGHT BINGE-WATCHING

Friday night is the waiting room on the journey to the weekend. What you want is a wine that is in tune with your euphoria at having two whole days off, and isn't going to look at you reproachfully if you slug it back without really concentrating. All of this means simple and pleasurable rather than complex. Big, cheerful flavors are in, elegance and restraint are out.

Most wines like this come from places with lots of sun. Australia, California, and Chile are great at making these luscious, up-front wines, jammed with fruit, and they do it, moreover, at good prices. Since the wine is to be drunk mostly on its own, we're looking for wines that don't have tons and tons of "structure"—which means soft tannins for red wines and low acidity for whites. Tannin and acidity might be essential when matching

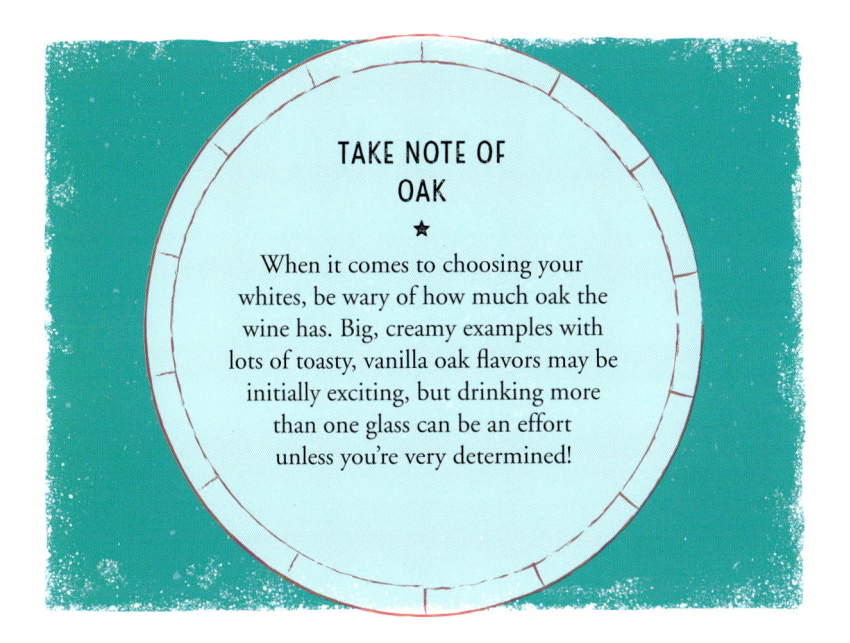

> **TAKE NOTE OF OAK**
>
> ★
>
> When it comes to choosing your whites, be wary of how much oak the wine has. Big, creamy examples with lots of toasty, vanilla oak flavors may be initially exciting, but drinking more than one glass can be an effort unless you're very determined!

with food (see pages 42–57), but can get in the way when you're just looking for a simple glass of glug.

So for whites I would rule out high-acid varieties such as Sauvignon Blanc or Riesling. Instead, keep your eyes peeled for an Aussie or California Chardonnay or a South African Chenin Blanc, all of which will be as comforting and generous as your sofa's cushions.

For reds I'd steer clear of Cabernet Sauvignon, unless you have a particular favorite. "Cab" is a great grape, but it can have quite strong tannins, which aren't what you want when you're drinking a wine on its own. Go for naturally softer varieties like Shiraz/Syrah or Merlot.

Australia, again, is great for the former, but California, South Africa, and Chile are all doing great Syrah, too. For juicy Merlot, California is your safest bet. If you want something a little bit different, Argentina does some great swigging Malbec. If you simply have to have a Cabernet, go for Chile, which makes the softest, most approachable examples in the world.

friday night binge-watching

BEFORE AND AFTER DINNER

There are times before a meal when only a beer, martini, or stiff gin and tonic will do. But wine has star performers of its own.

Fino sherry, for instance, is nutty, tangy, intense, and absolutely bone-dry. Served chilled right down, I have yet to find a better match with predinner nibbles, and its lack of bubbles means you won't get bloated.

Talking of bubbles, what about champagne? Yes, it's a bit of a splurge, but cheaper sparkling wines like prosecco can do a reasonable job for less-special occasions—just make sure the wine is dry (or *brut* in French).

In summer a fresh, chilled white like a Riesling, Muscadet, or Sauvignon Blanc can be a refreshing intro to a meal, and has the advantage that it might well carry on happily with the food.

But I'd think twice before serving red wine before a meal. It can swamp your taste buds and whatever food is coming next. If you've got fairly flavorsome nibbles, a chilled dry amontillado sherry (rich, nutty, and tangy) is a classy option and won't spoil if you leave it open for a few weeks.

When it comes to choosing drinks for after a meal, you're looking at the opposite end of the flavor spectrum. Instead of dry and light wines, you're looking for sweet and rich—the equivalent of a soothing massage for your palate rather than pre-exercise stretches.

Oddly, some sweet wines can

DECANTING VINTAGE PORT
★

Vintage port is unfiltered, which means it has sediment in the bottle. To make sure this doesn't end up in your glass, you need to decant it. Leave the bottle to stand upright for a couple of hours to settle, then pour it very slowly into your decanter (or pitcher if you don't have one) a few hours before serving.

If you carry out this process over a candle or a light, you'll see the sediment start to gather at the shoulder of the bottle. When it's about to come over the shoulder and down the bottle's neck, stop. There'll probably be half an inch of port left in the bottle.

be almost too unctuous on their own, and need food—Sauternes is a case in point—but those with a naturally higher acidity such as German Riesling (Auslese or Trockenbeerenauslese {TBA}) are a good post-dinner treat.

Sweeter champagnes used to be hugely popular and are a suitably decadent end to an evening. But for most people the traditional last wine is a fortified: Madeira or sweet sherry (occasionally) or, more likely, a port. Ruby ports are the cheapest, but late-bottled vintages (L.B.V.) or tawnies give far more bang for your buck. Vintage port has undeniable class, but you need to know how to serve it (see above).

BARBECUE

The joy of most barbecues is their ad hoc nature—the relaxed feel of eating outside coupled with the chance to pile your plate high with all manner of foods. Both these elements will have an effect on the sort of wines you serve.

Let's take the "outdoors" bit first. What you want to drink if it's lunchtime and baking hot is different if it's seven in the evening and starting to cool off. Generally speaking, the hotter it is, the more attractive those light, refreshing white wines are going to seem.

So, if it's hot, look for bottles with plenty of fresh, spritzy acidity and not too much weight. Sauvignon Blanc is perfect, but so are Riesling, Muscadet, Soave, and unoaked Chardonnay. Just make sure you keep the bottles chilled.

These wines might not, strictly speaking, be perfect with the food, but they're perfect with the weather, and if you're serving salads, fish, or even chicken, they'll be OK.

As it cools off, red wines become more appealing, but it's unlikely ever to be cold enough for you to want huge reds with high alcohol and enormous mouth-drying tannins. Soft, luscious examples like Pinot Noir, Merlot, Chilean Carmenère, Rioja, Malbec, or Beaujolais are a good bet.

When it comes to matching with food, the eclectic nature of barbecue food makes an across-the-board match impossible. How can you get a single wine that will go with salad, chicken, pork chops, and steak?

This is why I wouldn't get too hung up about the wine and go for general rather than perfect matches. Mid-weight whites like Chenin Blanc, lightly oaked Chardonnay, Sémillon, Verdelho, or Viognier will work with everything but red meat. Mid-weight reds like Merlot, Argentine Malbec, Barbera, Grenache, southern French reds like Fitou, and young Rioja will be fine with pork, lamb, burgers, and even chicken up to a point. Zinfandel can be good, too, but it tends to be high in alcohol, so be careful if it's really hot.

Save the bigger, more "classic" reds for formal sit-down barbies, but even here I wouldn't go for anything too heavy unless it's fairly cool (or you're eating indoors).

16 pick right at home

COOKOUT TIPS

> Rosé wines, chilled right down, are perfect for the barbecue. Served cool, they are as crisp and refreshing as whites, but take on more weight as they warm up. Make sure they're dry.
> Some reds work really well chilled, too. Beaujolais, Loire Cabernet Francs, and Pinot Noirs are the classics.
> Alcohol with sun is a potent combination, so make sure you offer plenty of water. One glass of water for every glass of wine is about right.
> No need to spend big with barbies. The food is fun rather than classy, so budget accordingly—maybe just above your Wednesday night wine.

SUNDAY LUNCH

Sunday lunch is something of an in-between occasion. It's several steps up from the rush of midweek dinners, yet not as involved as a gastronomic blowout; the sort of meal that's a bit special, but not one you want to spend all day preparing.

Treating yourself with your choice of wine is one of the best ways to instantly generate a feeling of indulgence without any extra effort. I'm not suggesting you break the bank, but adding 50 percent to what you spend on a midweek bottle should make a significant difference to the quality.

Trading up is a particularly good idea with your Sunday lunch, because good wine shows particularly well when it's not fighting with dozens of huge flavors in the food. And, since typical Sunday lunches are roasts, then by their very nature—solid food that's simply served—they are fabulous for showing off a better-than-average bottle of wine.

LAMB

A juicy Spanish red from Rioja or Ribera is a classic combination, but the meat also goes well with red Bordeaux and Chianti. Again, paying a bit extra makes a big difference to the quality, so if you go Spanish look for *reserva* (or *crianza*) on the label to prove it's had a bit of love and attention.

CHICKEN AND PORK

Both of these are at home with either whites or light reds. For characterful whites, it's hard to beat Burgundy with its fascinating Chardonnays. The trouble is that the best wines are expensive, while the cheaper versions are rarely worth the money. So, if in doubt, spend a bit extra!

Australia and New Zealand both do some flavorsome Chardonnays, while Aussie Semillon is a good mouth-filler. If you're having white meat with creamy sauce, a bottle of Riesling is a great and characterful match. Australia's are the best of the New World, but my preference would be for a dry wine from Alsace if you can find it.

If you fancy a red wine with your white meat, Pinot Noirs like red Burgundy (same rules on price apply as for white Burgundies) are best. Otherwise, go for a light- to mid-weight Italian red like Valpolicella or Chianti.

BEEF

Here you need the heavyweights of the wine world: big Cabernet Sauvignons and Merlots or Syrah/Shirazes. A decent bottle of Bordeaux has the most class, but if you want plenty of sweet blackcurrant fruit, Cabs from Chile, Argentina, South Africa, Australia, New Zealand, and California are all worth a look as well.

For Syrah/Shiraz, Australia and the Rhône are classic areas. The former tend to be bigger and richer, and work well if you're having intensely flavored game such as venison.

VEGETARIAN DISHES

Vegetables are like any other ingredient—you need to think about the way they're cooked when debating what to drink.

Spring vegetables and salads—asparagus, peas, and fava (broad) beans, for example—work well with crisp, fresh, fruity whites, such as Sauvignon Blanc and Grüner Veltliner. Suss out dry rosés and medium-bodied southern French and Italian reds for Mediterranean vegetables, including tomatoes, peppers, eggplant (aubergine), and zucchini (courgette). A buttery Chardonnay goes particularly well with sweetcorn, squash, and pumpkin, but opt for a lighter version or a Pinot Noir when serving mushrooms. Warming winter vegetables (think onions, carrots, parsnips, and dark, leafy greens) are often served in hearty stew or soups, which tend to suit rustic reds. Bakes and wellingtons containing beans or cheese suit hearty reds, but check they are suitable for vegetarians.

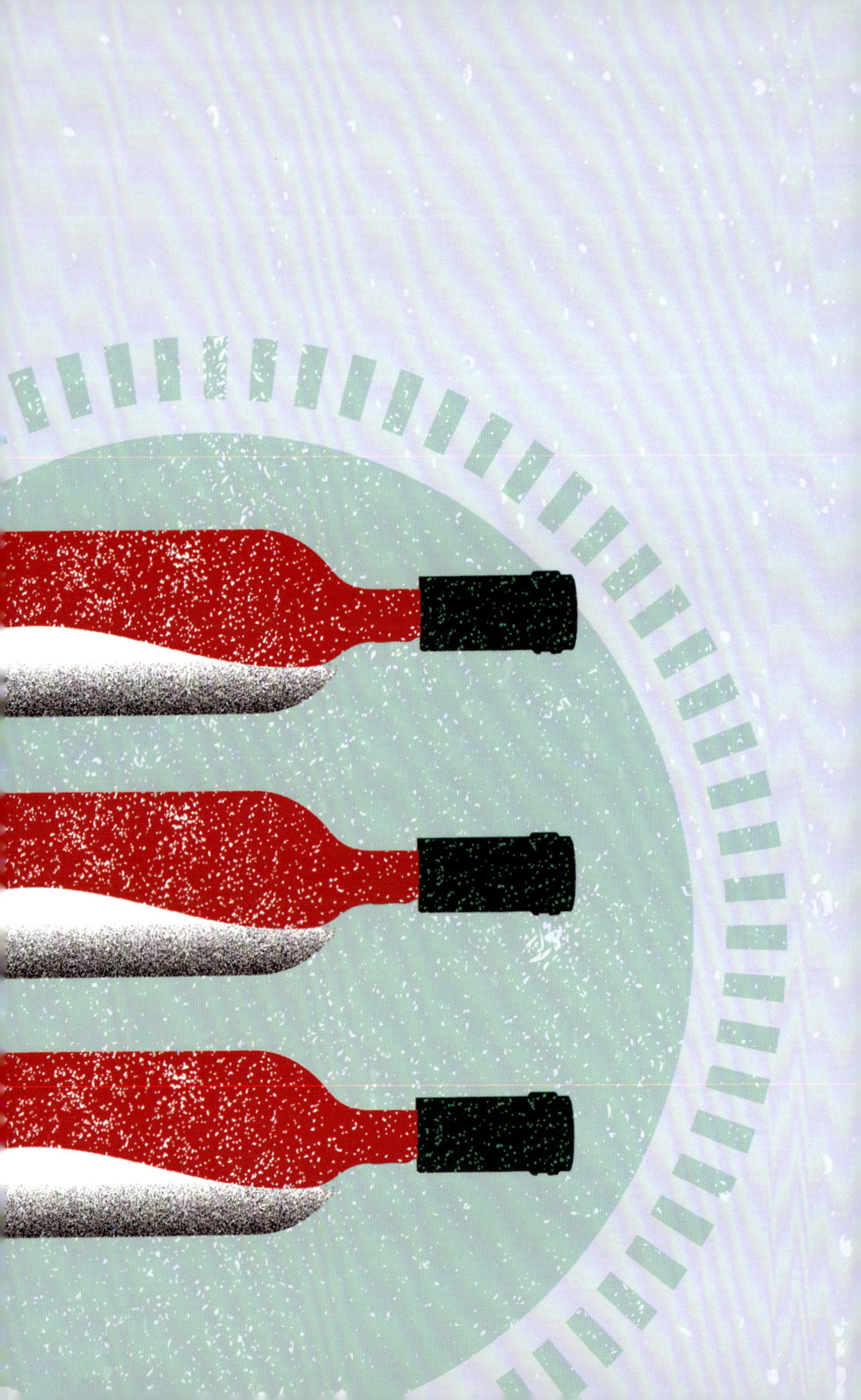

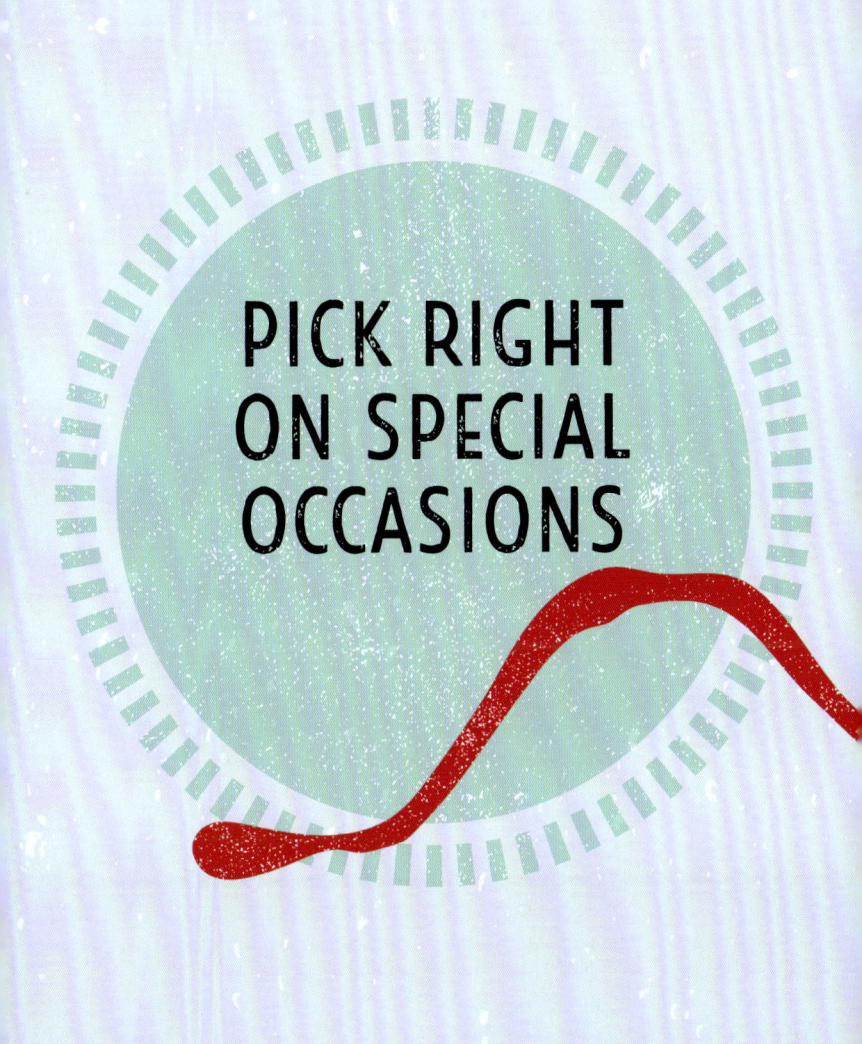

PICK RIGHT ON SPECIAL OCCASIONS

DINNER PARTY

There are different levels of dinner party—from the relatively informal "friends around" to the full-on five-star bash that takes weeks to plan and days to prepare. The way you approach your wine depends on how special the occasion is. What is true, though, is that these are occasions when going the extra few yards, not just in the level of the wines you serve, but also in the variety of bottles on offer, makes a big difference.

First, the groundwork—getting the quantity right. Each bottle holds six glasses of wine. So, if there are four of you, assume that you'll need one bottle for the starter and two bottles for the main course. Next, are you having sparkling wine to start? Dessert wine? Port? If so, the first two will need chilling, while the latter may need decanting (see page 15).

When it comes to the level of wines served, this too depends on the occasion. I wouldn't ever go below the Sunday lunch level. And if it's a long-lost cousin visiting from Australia, you'll need to show it's a big deal with at least one really good bottle.

The good news (for your budget at least) is that if you're serving two bottles of wine with the main course, they don't both have to be superstars. In fact, having one great wine and one good one can make for an interesting exercise in "compare and contrast!"

For sheer class, it's hard to beat the great areas of Europe: Bordeaux, Burgundy, Chianti, Rioja, and Barolo (for reds) and Burgundy, Alsace, and the Loire (for whites). At the mid- to upper price levels, the Rhône, south of France, Chile, South Africa, southern Italy, and Australia all offer good value for money. I'd suggest one wine from each group.

WHAT TO TAKE

★

As a guest, start by asking your host what the meal is and taking something that might go with that. If they prefer not to tell you, assume that your wine won't be drunk that night and take them something of a quality that will at least do for a Sunday lunch (see pages 18–19).

Don't, whatever you do, take them something cheap and cheerful—whichever way you look at it, it's an insult. Better to take good chocolate than bad wine.

DINNER PARTY TIPS

> Serve your wines properly, which means chilled for whites and (if possible) decanted for reds.
> A bottle of sparkling wine to start with really gives an evening a lift and needn't be that expensive.
> When it comes to buying a bottle of something a bit special, try your local independent merchant rather than a supermarket. Tell them your budget and the occasion, and they'll probably pick out something fantastic for your dinner party.

WEDDINGS AND ANNIVERSARIES

Making a success of big events like these depends on two things: one, making sure that there's plenty of fizz; and two, that it's drinkable. This might sound like an exercise in the blindingly obvious, but it's amazing how many big events run out of reception sparkly after one glass, or expect their guests to toast the happy couple with something resembling fizzy floor-cleaner.

For the reception sparkler, you'll need something dry. If you go for champagne, look for *brut* (dry) or "extra dry" (just off-dry, confusingly) on the label. For after-dinner toasts, find something sweeter (*demi-sec*).

But unless you have a huge budget, I'd skip champagne. Yes, at the top level it's gorgeous, but at the lower end it's rotten value for money. Instead, try a premium cava or prosecco (a good one should still be less than half the price of champers), a fizz from the New World (New Zealand is a particularly good source), or a bubbly from France that's not from the hallowed Champagne region. *Crémants* from Burgundy, Alsace, and the Loire can be really good value for money.

A bit of market research is essential before buying big volumes, so make sure you try out all your prospective candidates. Tasting a dozen bottles with friends will be a fun night as well as worthwhile, believe me! When it comes to the wines for the meal, follow the same rules as for a Wednesday night and look for something that's soft, fun, and palate-friendly at a decent price. Chile, South Africa, south of France, southern Italy, and Argentina should all be on your radar.

Incidentally, if you're buying half a dozen cases of wine from one place, you should be able to negotiate yourself a pretty good discount—don't be afraid to try!

As for amounts, I'd allow half a bottle of sparkly per person and about the same of wine for the meal. Thus, for ninety guests, I'd buy 48 bottles of sparkling wine, maybe 36 bottles of red wine, and 24 of white.

Remember, no guest ever complained that there was too much wine, but they'll always grumble if there's too little.

IT MAKES SENSE TO DO IT YOURSELF

Supplying your own wine allows you to serve decent wine at a good price. Most venues mark up a wine by around 300 percent, but many charge a relatively small corkage fee for every bottle of yours that they have to open.

Rates vary, but even if you pay $12 (£10) a bottle, which might be double with corkage, you'll have a decent wine that's likely to be better than anything you can get on the wine list at the same price.

Corkage on sparkling wines is higher (usually twice the price), but the principle is the same. Do the math. You know it makes sense.

THE HOLIDAYS

With its riot of flavors and textures at the dinner table, the holidays are a time of exuberant abandon for your palate, which can be both a liberation and a curse when it comes to getting the right wines.

Pre-dinner fizz

If you really want to start the day with a bang, champagne or Buck's Fizz is a decadent start. For Buck's Fizz (champagne and orange juice), I'd just use any cheap sparkling wine. If you want your fizz neat, go for a non-vintage *brut* (dry) champagne or a New World sparkler (New Zealand's are good).

The dinner

If you're looking for the perfect "silver bullet" match for turkey with cranberry sauce, stuffing, and potatoes, forget it—there isn't one. The flavors (from sweet through savory) are just too diverse—and that's before you get into the optional extras.

Now, while that does mean that you can't plonk one bottle down on the table and expect it to do the whole meal, it also takes a lot of the pressure off the event. Since there's no such thing as an across-the-board match, you can serve your guests a number of different wines and let them make up their own minds about which they like the most. It gives you the chance to stir up a bit of debate when it comes to your selection.

Your best bets are at the light-to-medium end of the red spectrum and the rounded, flavorful end of the white. For classic red matches, we're looking at Pinot Noir/red Burgundy or a decent Beaujolais. If you try the latter (and it's a good match), try to get one of the "*crus*" from a village like Fleurie, Morgon, or Moulin-à-Vent. They'll have more concentration than just an ordinary Beaujolais.

You can also try lighter examples of Syrah/Shiraz. Probably not the powerful, inky versions from Australia, but more flowery, red-fruited examples from northern Rhône areas like Crozes-Hermitage or New World countries like Chile or South Africa can work nicely.

The uniquely American Zinfandel has the advantage of working pretty well with holiday fare. White Zinfandel will be too sweet, but the red will be fine. Just keep an eye on the alcohol levels. Since it frequently gets up around the 15 percent mark, don't expect sparkling post-dinner conversation if your guests have gone through a couple of bottles.

When it comes to whites, look for dry Riesling or Chardonnay. Fumé Blanc (an all-American version of Sauvignon Blanc) is a slightly lighter alternative.

TOP TIPS

> Don't get hung up on looking for the perfect wine match. Just make sure there are plenty of different bottles on offer.
> Given the rustic nature of the food, you don't need to go crazy on the quality. Sunday lunch level and slightly above is fine.
> Don't serve heavy wines unless you want your guests to fall asleep.
> Branch out a bit with your wine choices.

PARTY

To an extent, the best advice I can give for party wine is "make sure there's plenty of it." But I'd also like to qualify that statement by adding, "do make sure that it's drinkable." If you stock up with the cheapest stuff you can find, it'll probably still be there afterward and you'll have to drink it yourself over the next months. If in doubt, find the cheapest wine on offer, add 30 percent to the price, and don't buy anything below that.

Before charging off to the store, it's worth asking yourself the following questions to help you formulate a plan of attack.
> Is it the sort of occasion (say, moving into a new house) that might require some form of celebratory fizz?
> Is it a hot-weather party or a cool-weather one? In hot weather people tend to drink more white wine.
> Will there be food or not—and what sort will it be? A big vat of lasagne, for instance, is likely to see people reaching for the reds.
> How many people are coming and how much are they likely to drink? Unless your friends are dipsomaniacs, allowing one bottle per person will be more than enough, particularly given that many, if you are lucky, may bring a bottle.

Your answers should allow you to figure out how much wine you want to buy and, roughly, the proportion of red and white you'll need. After that, it's on to the styles.

Essentially, for parties, you want wine that is similar to that for Friday night (see pages 12–13): simple, undemanding, with a decent whack of fruit, and not too high in either tannin or acidity. Something, in short, that slips down easily.

For whites, New World Chardonnay is a good choice (even with a fair bit of oak), as are South African Chenin Blanc, California Fumé Blanc, Aussie Semillon, or Riesling and Viognier.

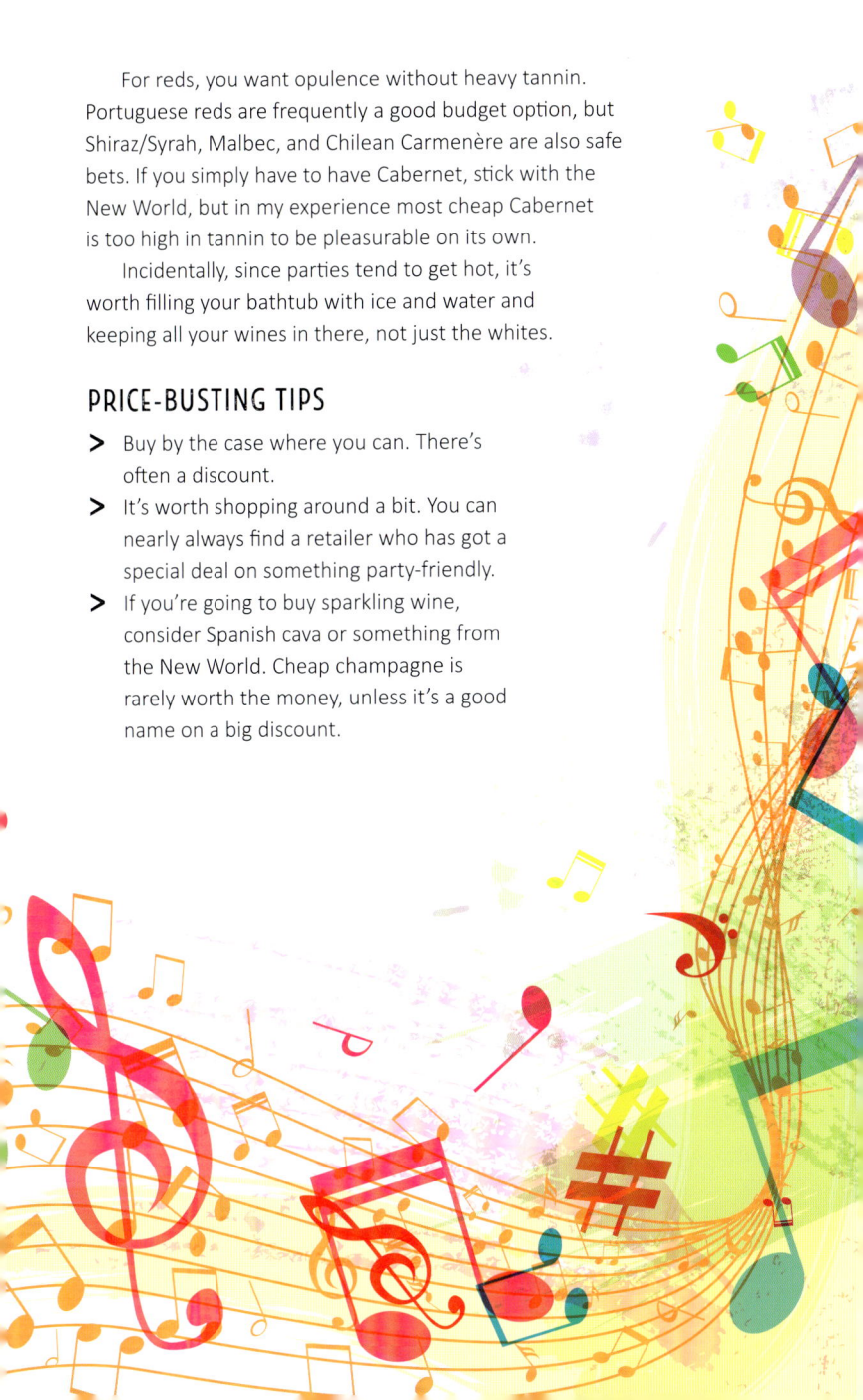

For reds, you want opulence without heavy tannin. Portuguese reds are frequently a good budget option, but Shiraz/Syrah, Malbec, and Chilean Carmenère are also safe bets. If you simply have to have Cabernet, stick with the New World, but in my experience most cheap Cabernet is too high in tannin to be pleasurable on its own.

Incidentally, since parties tend to get hot, it's worth filling your bathtub with ice and water and keeping all your wines in there, not just the whites.

PRICE-BUSTING TIPS

> Buy by the case where you can. There's often a discount.
> It's worth shopping around a bit. You can nearly always find a retailer who has got a special deal on something party-friendly.
> If you're going to buy sparkling wine, consider Spanish cava or something from the New World. Cheap champagne is rarely worth the money, unless it's a good name on a big discount.

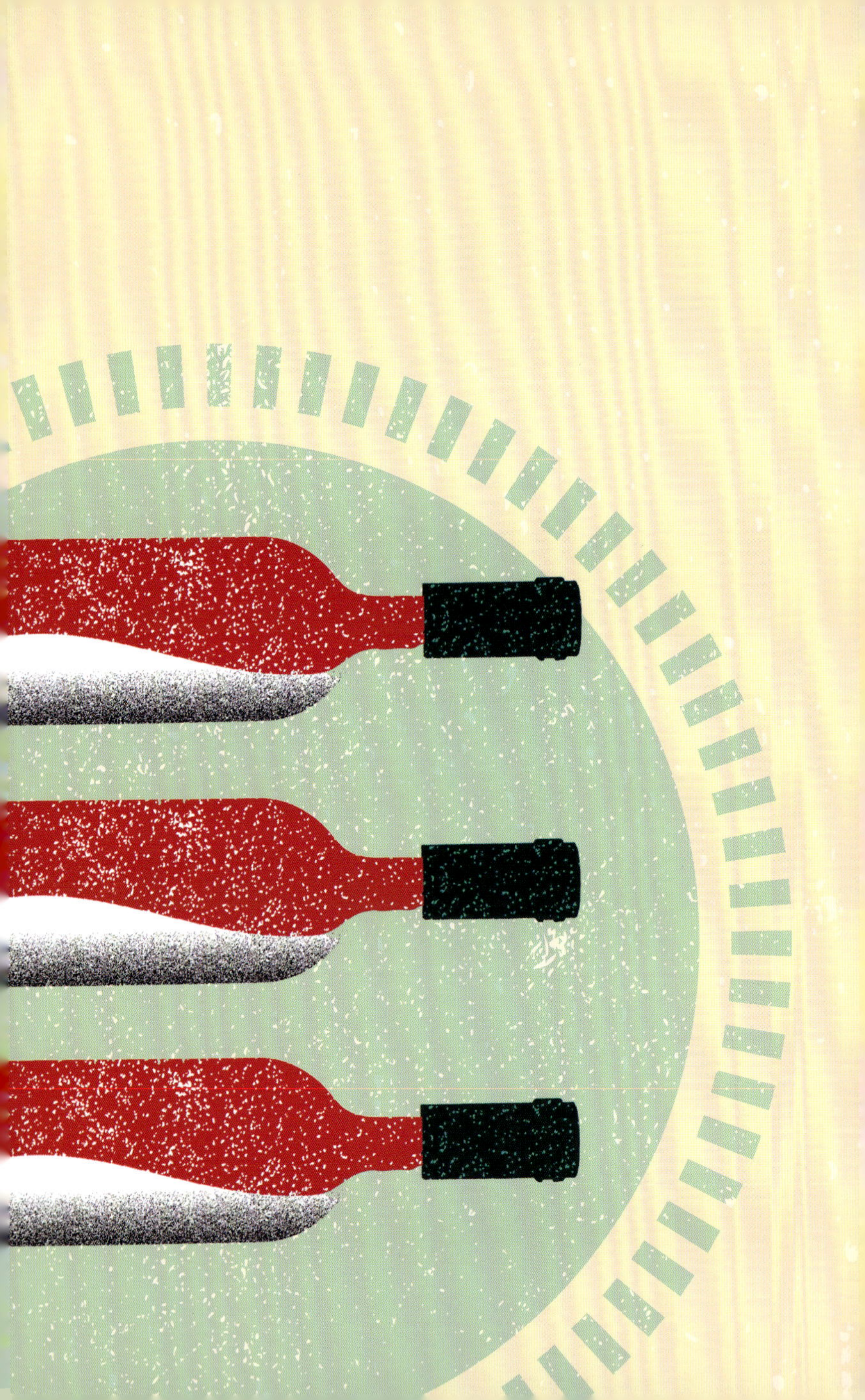

PICK RIGHT OUT AND ABOUT

BARS AND PUBS

Boy, does this chapter have to cover a lot of ground! From swanky specialist vinothèques through glitzy bars to spit-and-sawdust pubs, you can get wine of some description in all of them, but how to avoid the rubbish?

Three basic rules

Take your cue from your surroundings. If it looks like the sort of place that serves six bottles of wine a year, the stock isn't likely to be either fresh or well kept, so drink something else.

If it's a really good specialized wine bar, the staff may have a few good recommendations of their own, so ask them.

Keep your eyes peeled for promotions. Rather than ones that are encouraging you to drink more (a good way of offloading cheap stock), go for promos that cut the price of an expensive wine to an affordable level.

Choosing a wine

> Generally, you want soft, fruit-forward, easy-to-drink wines rather than "classical" clarets and Chablis, which have more tannins (red) or acidity (white).

> I usually find white wines easier to drink in bars, provided they're not too warm. New World Chardonnay is a decent white choice, though some cheap examples can be a bit sickly, so avoid the bottom-end stuff. Sauvignon Blanc is a refreshing palate-sharpener on hot days, but is usually too acidic for more than one glass, so be prepared to switch unless you want indigestion. White Bordeaux (a blend of Sauvignon Blanc and Sémillon) can be easier on the stomach.

> Great by-the-glass whites for me are Riesling, Gewürztraminer, or Tokay Pinot Gris from Alsace, but they aren't always easy to find. Pinot Grigio, the Italian version of Pinot Gris, is more common (and varies from fun and sprightly to thin and nasty, so pick with care).

> If you go the red route, New World Cabernets, Merlots, and Shirazes are soft and juicy, but also get rather heavy for repeat drinking. Spanish reds (Tempranillo or Monastrell) or lighter Italians (Barbera and Valpolicella) are a good alternative. As, believe it or not, is Beaujolais if it's a decent example.

DATE NIGHT

This is arguably the hardest of all the categories to get right, since its success depends as much on your attitude as the actual wine you order. That's why it's really important to talk to your date about their preferences before you order. They may well not have strong opinions either way, but it really helps to know if, say, they don't like Chardonnay.

Once likes and dislikes have been safely worked out, it's on to deciding how much to spend. This, in a sense, is the tough bit. After all, we are all capable of ordering a less-than-brilliant wine from a wine list, but if we do it by choosing the cheapest on the menu, then we can fairly stand accused of being stingy.

So, without bankrupting yourself, I suggest trawling the wine list at a slightly higher level than you would usually. As a general rule of thumb, take the price of the house wine, double it, and add a bit. At that level you should still be some way below the super-expensive wines (which could lay you open to charges of being flash), but safely into some pretty good stuff.

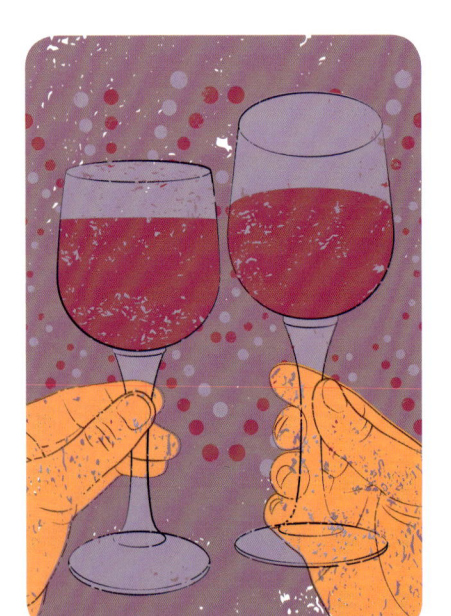

At this price, the wine ought to speak for itself, but even if it's a bit disappointing, you can hardly be blamed. Having sought a consensus on what to choose at the start and then spent a decent amount on trying to achieve it, you can squarely be said to have done your best. And on a hot date you can't really ask for anything more.

IF YOU'RE GOING CLASSICAL, DO YOUR HOMEWORK

★

Ordering Old World classics like Bordeaux and Burgundy can impress, but it helps to know which vintages or producers to go for—and you don't want to be lost in silent contemplation for ten minutes at the start of your date.

If you know the restaurant you're going to beforehand, check out their wine list and do some research. (Heck, you can even take the list into a local wine store and ask them for help!) Either way, a bit of forward planning can help.

DO

> Start your date by ordering a couple of glasses of good champagne. It instantly gives the evening that special feel—and because of its bubbles, fizz lowers inhibitions more quickly, helping conversation flow.
> Consider more expensive wines from cheaper countries—they tend to be soft, rich, and flavorful. While not always a classic match, they give good bang for your buck and are very likeable.

DON'T

> Buy the cheapest wines on offer, even if your date isn't a wine buff.
> Spend ages agonizing over the wine list unless your date is a wine lover.

VISITING THE IN-LAWS

Forget Olympic finals or politicians being grilled by an interviewer on live TV, taking a bottle of wine to your in-laws is the definition of having to perform under pressure! Especially, I might add, if it's one of your early visits. And the bad news is that there's no such thing as a blanket recommendation, either. For instance, you might think a classy bottle of white Burgundy is the perfect present—but it's no good if they only drink red wine from the New World.

Much of the success of an in-laws' wine gift depends on your research, which, if you don't know their personal preferences, means grilling your other half. The information could be as specific as "they like 1990 St-Émilions" or as general as "they don't like white wine." Either way, establish some guidelines before venturing out with your checkbook.

If they have some favorites, then great, you know where to start looking. If they don't have any preferences or don't drink that much wine, then your job is both harder (because the potential choice is vast) but also easier, because, provided what you bring is better than what they're used to, they'll be happy.

For the uncritical family, wines that deliver plenty of bang for your buck are a good bet. A $15 (£12) red from Chile, Argentina, or South Africa will get you a lot of rich, silky flavors. Whites are harder, but a creamy Australian Chardonnay usually does the trick, as does a waftingly aromatic New Zealand Sauvignon Blanc.

For in-laws who drink wine more regularly but have no great expertise, the key is to find a wine style they like, then consider spending up to twice as much as they usually would in order to get a wine that is guaranteed to deliver. In this, you'll increase your chances of success if you avoid the supermarket in favor of somewhere with experienced staff who can give you decent recommendations.

When it comes to buying for wine-literate in-laws, there's only one option: go to a specialized merchant and ask them to recommend something in the relevant style. There are two advantages to this: one, it takes the pressure off you (beyond signing the check) and, two, a good merchant should be able to give you a bit of interesting background info about the wine.

Showing that you've gone out of your way to pick up something a bit interesting is worth many cred points and, even in the unlikely event that the wine doesn't hit the spot, you'll still come out of it well.

PICNICS

I always think that picnics are a bit like barbecues but without the chaos. For starters, the food will be limited, and so will the amount of wines you can take with you, which makes things a lot easier. Now there are occasions when only a good, chilled six-pack of beer will do, but equally there are times when wine is not just a good alternative, but even the best choice.

If it's a special occasion (say, a birthday treat), then why not push the boat out with a big-name champagne to match? You don't need to go the whole hog and buy vintage champers, but a non-vintage from a top house can really make it an occasion to remember. New Zealand makes some great sparkling wine or, at the lower end, cava from Spain is a fair alternative.

For whites, I'd keep it light and zippy—Sauvignon Blancs, Riesling, Soave, and the Spanish Albariño are refreshing when drunk outside in the sun. For reds, avoid the heavy stuff. Beaujolais is good chilled and juicily appealing, but young Rioja and Pinot Noir work well, too. As with barbecue wines, rosé is a great (and underrated) alternative.

PICNIC EXTRAS

★

These can make all the difference. For instance, you absolutely have to have a cool box. There's nothing worse than lukewarm champers or white wine. A chiller sleeve to put around open bottles isn't a bad idea, either—even for reds if it's really hot.

As for glasses, if you don't have a five-star hamper that will allow you to transport "proper" glasses safely, see if you can get some non-glass wine glasses. There's something incredibly depressing about drinking wine out of plastic stacker-cups, and even plastic wine glasses are a big improvement.

GIFTS

As with buying for the in-laws, gifting success depends not so much on spending a fortune as getting a wine that will suit the people you're buying it for. For instance, while a wine buff isn't likely to be impressed with a $7 (£5) bottle of easy-drinking plonk, there's absolutely no point in buying *cru classé* Bordeaux for someone who only ever drinks wine on the sofa in front of the TV. Likewise, if you're taking wine along to a barbecue, you need to make a decision as to whether the bottle is intended for consumption at the event or as a separate present to be drunk later by the hosts.

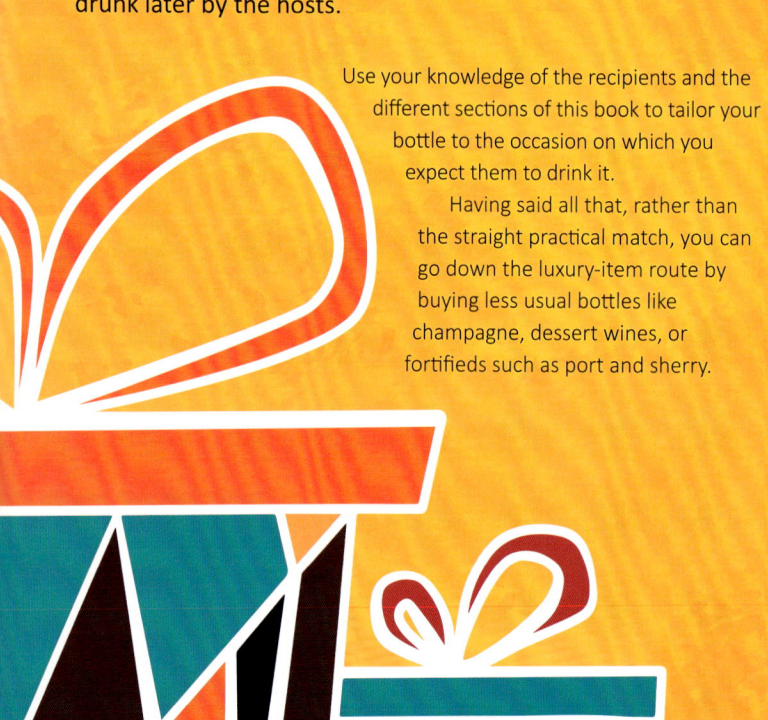

Use your knowledge of the recipients and the different sections of this book to tailor your bottle to the occasion on which you expect them to drink it.

Having said all that, rather than the straight practical match, you can go down the luxury-item route by buying less usual bottles like champagne, dessert wines, or fortifieds such as port and sherry.

These make great gifts because people often like them, but tend not to buy them for themselves regularly. The inclusion of these luxury wines turns any meal into an occasion, and they'll probably think of you when they pop the cork.

The best value for money comes from non-vintage champagne from a good name and late-bottled vintage port, which gives real after-dinner class without breaking the bank. Dessert wines—even Sauternes, the pinnacle—are well priced, considering how much work has to go into their production.

BOTTLE-BUYING TIPS

★

Bottles of port and champagne often come in their own carton, which makes them both easier to wrap and also better-looking than a bottle on its own. The same goes for spirits.

OK, it's not wine, but if you do decide to buy a bottle of spirits for someone, make sure that they like the style you're buying. A fan of sweet bourbon or cognac, for instance, is unlikely to enjoy a pungent, salty Islay whiskey—and vice versa.

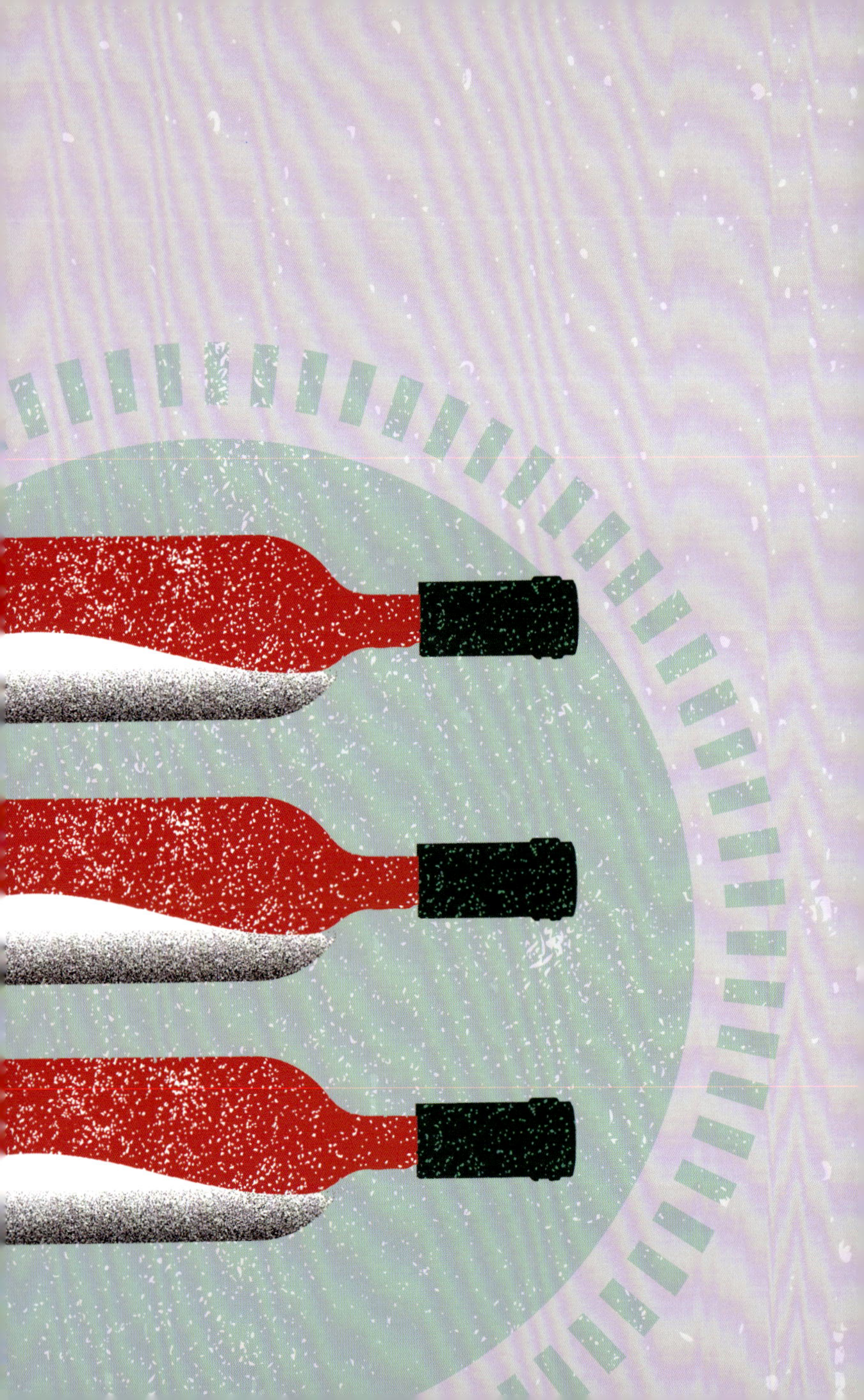

PICK RIGHT EATING OUT

NAVIGATING THE WINE LIST

It's perhaps not surprising that so many people find ordering wine in a restaurant a somewhat intimidating experience. But fortunately, the nightmare scenario of giant, incomprehensible wine list and snooty waiter happens less and less. Besides, if you follow the suggestions below, you'll stack the odds in your favor.

ASK THE WAITER
Far from being aloof, good wine waiters are usually only too pleased to help if you ask them for their suggestions. Tell them what you're all eating, what you generally like, and give them a price, and see what they come up with. Remember, it's their job! Just don't let them talk you into spending lots more than you want to.

LOOK FOR FAMILIAR FACES
If the restaurant doesn't have a wine specialist and your waiter looks clueless, look for brand names, grape varieties, or even countries that you like to drink on the list.

LOOK FOR CLUES ON THE WINE LIST
Nowadays, more and more wine lists have helpful descriptions of the flavors of the wines and even, in some cases, food-matching suggestions. Make the most of them.

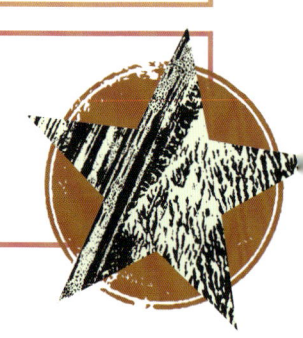

Try Before You Buy

A lot more restaurants are offering a decent selection of wines by the glass. If you try a selection with your appetizers (or before sitting down to eat), you'll be able to order bottles of your favorites with confidence. Single glasses can also be a fantastic way of keeping everyone happy if, for instance, your table of four orders red meat, white meat, vegetarian, and fish.

Avoid House Wines

The fact that house wines are the cheapest wines on the list doesn't, by any stretch, make them the best value. Typically, they're very, very cheap wines that have been hugely marked up, and I've rarely had even a halfway decent one. My benchmark would be to find the price of the house wine, add 50 percent, and start looking around that level.

Consider Safety First

If you are really struggling, certain wine styles make good "bale-outs"—not necessarily the best wines on the list, but solid and reliable. You won't go far wrong with Chilean Cabernet Sauvignon, California Merlot, Australian Chardonnay and Shiraz, Côtes du Rhône and New Zealand Sauvignon Blanc.

RESTAURANT MARK-UPS

These vary significantly from country to country and from eatery to eatery, but it's not unreasonable to expect a restaurant to triple or even quadruple the price of a wine that it's selling. In other words, if you habitually spend $10 (£8) on a bottle in the supermarket, you'll probably need to spend somewhere around $30 (£23) in a restaurant to get a comparable quality.

French

The French have a good claim to being the nation that gave great food to the world—and the same could be said when it comes to food and wine matching. The point about classic French food—unlike, say, Thai cuisine—is that because it's usually a fairly simple mix of ingredients, it is much easier for it to work in harmony with the wine, allowing both to shine at once. So, if you're ever going to splash out and spend big in a restaurant, this is the type of food with which to do it.

Obviously, in the suggestions below you could substitute, say, a California Cabernet for a Bordeaux. But I've stuck with French recommendations for two good reasons: French wines are generally very food-friendly; and your average French restaurant doesn't sell many wines from anywhere else!

Cheese and Pâté

Sweet Sauternes is a nailed-on classic match with both foie gras and Roquefort. For big, pungent (or smoky) cheeses, try Gewürztraminer from Alsace; for goats' cheese, Sauvignon Blanc is a classic (and classy) pairing. For Brie, Camembert, et al, white Burgundy is your man!

Shellfish

Oysters, shrimp, and *moules marinière* are absolutely at their best with zingy northern French whites like Chablis, Muscadet, and Sauvignon Blancs from the Loire.

Fish

Depends how it's cooked. For light, white fish, simply grilled, go with the flinty Sauvignon Blancs of Sancerre. Creamy sauces need something more substantial, like a white Burgundy or an Alsace Riesling. Meaty fish like monkfish can work with lighter reds such as Pinot Noir, decent Beaujolais, or Loire Cabernet Franc, as well as bigger Burgundies.

Casseroles

If there's wine in the sauce, then it's usually a good bet to use the same wine to accompany the casserole. Thus *coq au vin* and *boeuf bourguignon* both work with simple Burgundy, though the beef can also stand up to more substantial "comfort" wines, such as reds from the Rhône or the south of France. These dishes are simple, hearty food, so there's no need to spend big on the wine.

Chicken

In a salad, go with medium-bodied whites like Chablis or simple Burgundy. If it's roasted, find a good wine to show the chicken off, like a decent white Burgundy. If you're eating really good meat (say, *Poulet de Bresse*), red Burgundies work, too.

Beef

Depends a bit on how it's cooked and how good the cut is. Simple *steak frites* or more rustic joints are good either with cheap Bordeaux, Côtes du Rhône, or southern French reds. For top-quality filet, you want as good a wine as you can afford from Bordeaux, Burgundy, or classic Rhône areas like Côte Rôtie.

Game

Wines with a bit of age go best with any earthiness in the meat. Old Burgundies and Bordeaux are the classic matches.

France and Her Grape Varieties

WHITES

Loire/Sancerre: Sauvignon Blanc

Burgundy/Chablis: Chardonnay

Bordeaux/Sauternes: Sauvignon Blanc, Sémillon

Alsace: Riesling, Gewürztraminer, Pinot Gris

REDS

Bordeaux: Cabernet Sauvignon, Merlot, Cabernet Franc

Burgundy: Pinot Noir

Loire: Cabernet Franc

Beaujolais: Gamay

Northern Rhône (Côte Rôtie, Hermitage, etc.): Syrah/Shiraz

Châteauneuf-du-Pape: Grenache

Southern Rhône: Grenache, Shiraz, Mourvèdre

South of France: Grenache, Shiraz, Mourvèdre, Carignan

Italian

Italian food is justifiably among the most popular in the world. Its riot of colors and flavors, while retaining a suitably down-home feel, makes it a comfort food *par excellence*. It also has the singular advantage of being exceptionally wine-friendly, and most dishes work pretty well even if the wine match isn't perfect.

Pasta

It's all about the sauce. For seafood-based pasta (*vongole*, for example), you can go with just about any white that doesn't have lots of oak in it. Pinot Grigio will be fine, and so will Sauvignon Blanc, Chablis, or a lightish unoaked Chardonnay.

For aromatic, fragrant pasta sauces like pesto, you need a wine that isn't going to override the flavors. So, gently flavored Italian whites like Gavi or Soave can work, as can a Catarrato from Sicily. Outside Italy, Albariño, New World Riesling, a California Fumé Blanc, or mid-weight Chardonnays are fine.

For carbonara-type creamy sauces, you need a bit of structure to cut through the cream, but not too much fruit flavor, which may well clash. I'd recommend either the same sort of whites that match the pesto, or lighter reds like a Montepulciano, a cheap Bordeaux, or a Cabernet Franc from the Loire.

For meatier sauces like Bolognese, try the more heavyweight Italian portfolio of reds. Chianti is a good match, but I'd tend not to go for the cheapest. Good Chianti is great, but bad Chianti is a miserable experience, and spending a bit extra can make a big difference. Primitivo, from the south, is a good (and cheaper) alternative. Otherwise, any good, gutsy wine will do: Cabernet Sauvignon, Shiraz/Grenache blends (for example, from the southern Rhône Valley), Zinfandel, or Chilean Malbec. There's plenty of scope to choose your favorite—so go to it!

Pizza

With its massive mix of flavors, pizza is not an easy dish with which to get an absolutely perfect wine match. But then it's not that kind of food, so don't get hung up on it.

As a general rule, I'd favor wines with gentle rather than in-your-face flavors and of no more than medium weight. Even pizzas with lots of toppings tend not to be heavy, and a big, gutsy red wine will just stamp all over them.

For whites, a fresh, gentle Italian like Soave or an unoaked (and not over-fruited) Chardonnay work with most; for reds, Barbera and Montepulciano (from Italy), a southern French Roussillon or Fitou, or young Rioja are safe bets.

Other classics

Meaty lasagne and ravioli follow the same rules as Bolognese. For vegetarian versions, take the middle-of-the-road carbonara route. For veal, Soave is the lighter match. If it's served with a heavier sauce, try a Pinot Noir.

Chinese

As with all types of food, but especially Asian cuisines, there is no such thing as a "one bottle suits all" match for Chinese food. If you're in a decent-sized group, the best bet is to pick a selection of different bottles and try small amounts of different wines with each dish. If you're there as a couple, you'll probably have to resort to wines by the glass to get really good matches right the way through the meal.

The good news is that Chinese food does offer some genuinely good matches. Sauvignon Blanc (or champagne, if you're pushing the boat out) works with spring rolls and most seafood dishes, while lusher whites such as Chardonnay or Sémillon go with bigger-flavored chicken or pork dishes. Avoid anything that's searingly dry like Sancerre or Chablis—the sweetness of many Chinese dishes means the wine needs a little weight and lushness to it.

Another advantage of Chinese food is that, while it's usually heavily flavored, it's rarely actually hot, which means that red wines can play more of a part. As with the whites, the natural opulence in most New World wines means they usually make a better match.

Duck dishes are a classy partner with New Zealand Pinot Noir, while staples like beef and black bean sauce work well with bigger, sweeter New World wines such as Chilean Carmenère, California Merlot, and Aussie Shiraz.

The Gewürztraminer conundrum

Gewürztraminer is often touted as a good match with Chinese food, but its big, perfume-y flavors mean there's a possibility for hideous clashes, too. If in doubt, ask yourself whether the dish you've chosen will work with a side dish of litchi nuts, Gewürz's signature flavor. If it won't, stick with something more neutral, like Pinot Blanc or Chardonnay.

Japanese

With its purity of flavor, Japanese food can be one of the easiest of the Asian cuisines to match with wine, particularly if you're having sushi. The Japanese often drink sake with it, and for me the best match is the wine that is closest to sake stylistically: fino or manzanilla sherry, which have a delicate nutty, savory flavor profile and are bone-dry.

If you prefer non-fortified wine, the key is to get something dry and even a little austere. New World whites tend to have too much sweetness of fruit for sushi, so you're better sticking to Europe and "classic" regions like Sancerre, Chablis, or Burgundy, where the flavors tend to be flinty and chalky rather than actively fruity.

Meatier Japanese dishes obviously require red wine, but probably not anything too enormous, which might stamp all over the flavors. So, provided what you've ordered isn't too spicy, medium-bodied European reds are likely to be best. Young Bordeaux or Burgundies work well, and so can Chianti, Rioja, and Loire Cabernet Francs. If you want to go New World, a Kiwi or Aussie Pinot Noir (from Tasmania if they've got it) is probably your best bet.

For anything with lots of dominant strong flavors that are "difficult" for wine (for example, Japanese pickles, wasabi, and sesame oil), you might want to go native and try that sake!

And from left-field...

If you're looking for something a bit different, dry (*brut*) champagne can be an inspired match with sushi, particularly if it's a good, chalky *blanc de blancs* (i.e. made from Chardonnay). For teppanyaki, a rich, lush amontillado sherry can do the trick.

Thai

With its myriad spices and penchant for chilis, it's difficult to get a perfect wine match with Thai food. The most important thing is to gauge the heat of the dish. If you crunch down on a green chili, don't expect a glass of dry red wine to bring you much relief. In fact, very few Thai dishes suit red wine, so you can more or less head straight for the white section of the wine list, whatever you're eating.

The aromatic nature of Thai dishes is better suited to the flavor profiles of white grapes. Red wines either override the spices completely or, worse, clash with them. White wines have no tannin, which is a Good Thing (see Tannins and Heat, below). Not only that, but what they do have is higher acidity, which can help to freshen your palate and counteract any heat in the dish just as effectively as beer.

A little sweetness isn't by any means a bad thing either, so my top recommendations for Thai food would be German Riesling (Spätlese or Kabinett if they've got it), juicy, aromatic Sauvignon Blancs from New Zealand, and apricot-scented Viogniers. For hot, but light dishes (say pork, chilis, and basil), try a fresh, neutral Soave or Pinot Blanc. For meatier dishes, go with an Alsace Tokay Pinot Gris.

Tannins and heat

Tannins cause the mouth-drying sensation (you can feel it on your teeth) that you get in red wine. They're an essential part of red wine's makeup and, usually, an essential ingredient in providing "structure" when it comes to food matching. But because they dry your mouth out they exaggerate any heat in food, so be very wary about ordering red wine with spicy dishes. If you insist on red wine, make it a soft one like a Dolcetto, Beaujolais, or Valpolicella.

INDIAN

The biggest problem with matching wine to Indian food comes down to one word: heat. If you are addicted to searingly hot dishes such as vindaloo, you'll just have to accept that your choice is, to put it mildly, limited. But if you prefer creamier or more medium-spiced dishes, then things get easier.

If in doubt, the safest bet for Indian food is, as with Thai, to avoid red wines for reasons of tannin (see left) and look instead at whites. Since the dominant flavor in Indian cooking tends to be the sauce rather than the meat, it's perfectly possible to have red meat and white wine.

There are two routes to go down: either find a big, pungent white that will wrestle with the food head-on (like an Alsace Gewürztraminer), or go for something deliberately more neutral that acts as a "straight man" to the food and soothes the spiciness.

Lush Tokay Pinot Gris or Pinot Blancs (again from Alsace) are good bets here. For creamier dishes like korma or passander, Rieslings are a decent match. For fish curries, New Zealand or South African Sauvignon Blanc work OK. A personal favorite of mine is white Rhône, which can work with many dishes, though it's not always easy to find.

If you fancy something out of left-field, big dessert wines like Tokaj work surprisingly well. They have the lushness to stand up to even pretty hot dishes and plenty of cleansing acidity to keep the mouth fresh. You won't find many on restaurant lists, but if you're entertaining at home they could be a real crowd-pleaser.

MEXICAN

With its robust flavors and cheerful spices, there's absolutely no point in getting too fancy when it comes to matching wines with most Mexican food. So, unless you're going to a trendy CaliMex fusion restaurant (when things get kind of complicated and your best bet is probably to ask the waiter for help), the key is to keep it simple.

With chunky food like chicken burritos, you're looking at the soft, round, and full-flavored end of the spectrum rather than delicate and aromatic, which leads us once again to Chenin Blanc, Chardonnay, and/or Sémillon country. A bit of oak is fine, but remember that anything that's too heavily wooded will increase any heat in the dish.

For more esoteric seafood fare featuring, say, shrimp, limes, and chili, Rieslings can be an inspired match, and the acidity will freshen your palate as effectively as beer.

For chunky red meat dishes, you need a wine that is not overly tannic, so that the heat in the food isn't exaggerated. This rules out many Cabernet Sauvignons, but there are still plenty of wines out there with a bit of attitude that work well. Southern French reds (Rhône Villages, Languedoc) have a bit of grunt to them, as does Shiraz.

But, for me, the best match with tortillas, tacos, et al, is California's own Zinfandel. Big and lush, it manages to be spicy without being aggressive, and soft without being flabby. Well worth a look—just don't confuse it with white Zinfandel (a sweetish pink version, made from the same grape).

Middle Eastern

The classic Middle Eastern cuisine features a mixed selection of starters (mezze), followed by a more solid meat or fish course. And the good news is that, with its emphasis on natural ingredients and generally simple, fresh flavors, wine matching is something of a dream with most of them. Why? Because there aren't too many conflicting flavors to get in the way, and the dominant flavors, garlic and herbs, are far easier to match with wine than, say, chilis and lemon grass.

For mezze, a good, crisp white wine is the best starter. A modern white Rioja or Albariño from Galicia in Spain; a Loire Sauvignon Blanc; a herby Catarratto from Sicily; or a cool Soave—even a crisp Portuguese white—will all work with the majority of Greek and Lebanese starters.

You can even try an unoaked Chardonnay with lamb mezze if they've been drizzled with lemon juice, and the wine could then continue with the main course if you've ordered grilled fish.

For meats, again, the simplicity of flavors makes matching reds comparatively simple. Huge, sweetly fruited wines aren't likely to be a great match with savory, herby flavors, so you are probably best off looking in Europe or cooler areas of the New World. Bordeaux, Rioja, and Ribera del Duero are good, as are Chianti and Fitou. If you want an Aussie Cabernet, try Western Australia.

Go Native

There are plenty of good Greek wines appearing now, frequently blending unpronounceable native grapes like Agiorgitiko with more familiar ones like Cabernet Sauvignon. They're worth a look. And if you fancy an established classic, Château Musar from Lebanon has been a powerful, chunky, cult wine for decades.

Seafood

While with seafood it's broadly true that if you order anything chilled and white you won't be too far off the mark, there are some matches that are indisputably better than others. After all, there's a world of difference between the delicate flavors of, say, lemon sole and meaty fish like tuna.

Crawfish, shrimp, and the like work really well with very delicate, fresh whites like Muscadet. For simple fish dishes (without strong sauces), you'll never be far wrong with a Sauvignon Blanc, particularly if you shell out a bit extra and get one from Sancerre in France. Fresh and steely, it's a classic match with the likes of sole.

For meatier fish like halibut and sea bass, you need a bigger wine, and, while you can still get away with a good Sauvignon Blanc, a lightly oaked (or unoaked) Chardonnay, like a Chablis, will carry the day nine times out of ten.

For seriously meaty fish like tuna, you want biggish (lightly or unoaked) Chardonnays from countries like Australia or South Africa or even light reds like Pinot Noir, Beaujolais, or Cabernet Franc.

SEAFOOD TIPS

★

If you're having simple crawfish as an appetizer, why not accompany them with a chilled glass of fino sherry? Elegant, classical—and very Spanish!

If you have to pick one "bale-out" wine to go with a variety of different dishes, Chablis is probably your best bet. It's fresh, but with reasonable body.

Fusion Food

This is some of the toughest food to get a good wine match with. For starters, it's non-traditional, so there are no tried-and-tested classic matches; and its tendency to throw unusual tastes together can make it hard to predict flavors just by looking at the menu.

The good news is that, since most fusion restaurants tend to be fairly serious establishments, they often have a knowledgeable wine waiter—so don't be shy about asking him or her for advice.

Although you can usually come up with something reasonable by a process of elimination (if a dish is Thai-influenced, you can follow the rules on page 52, for example), for me, the key is not to get too hung up on the search for perfection.

Fusion food is fun, and you need to buy into that spirit of adventure. Be like the chef and try stuff out. Think laterally, throw out the rule book—and enjoy yourself!

ADVENTURER'S TIP
★

A fair few fusion dishes mix big flavors to make an even bigger combination. This can be too much for most wine styles, but sherry (yes, sherry) can work really well for these power-plates if you get the right style. Try manzanilla or fino with fish, amontillado with game, and oloroso with red meat.

STORING

When it comes to eating at home, buying the right wine in the first place is only part of achieving vinous nirvana; you need to treat your new purchase right if you're to get the most out of it.

OK, if you're planning to drink a wine within a week of purchase, then short of storing it in the oven or the freezer it doesn't matter much where you keep it. But any longer than that and it helps to bear the following in mind. Trust me, you'll save yourself a lot of ruined bottles.

Where to store it

Wine likes the sort of conditions that humans don't: cool, humid, and dark. The ideal temperature is about 50°F (10°C). If you have a cellar, perfect. If not, find a place that checks as many of the boxes as possible.

But (and it's a big but) just bear in mind that big temperature swings are an absolute no-no.

So, storing it in the garage or under the eaves of the house is out. Better to find somewhere that's dark and a bit too warm but constant—maybe a hallway or under the stairs where it's a steady 60°F (15°C)—than somewhere the temperature fluctuates wildly.

How to store it

In a word: horizontally. The lapping of the wine stops the cork from drying out.

How long to store it for

This depends on the environment and the wine itself. The warmer it is, the faster a wine ages, so a bottle that lasts years in a cellar will probably be shot after six months in your living room.

Then there's the question of whether your wine will improve with age, even if it's well kept. Most modern wines are ready to drink straight away and are likely to deteriorate with time. So, if in doubt, pull the cork!
NOTE Fine wines are an exception to this "drink young" rule.

SERVING

Serving wine correctly is easy—and doing it well can have a big impact on what it tastes like. The following three variables make a big difference.

Glass Size

Essentially, the bigger the better, since you can slosh the wine around and release its aromas. Titchy wine glasses should be thrown out.

Temperature

There's a tendency to overchill whites and serve reds too warm. Bad news. White that's too cold loses all its flavors, while a hot red tastes unbalanced and alcoholic. Most fridges are colder than the ideal white wine temperature, so take the bottle out half an hour or so before serving. Conversely, the idea of "reds at room temperature" was established long before central heating, so you might want to leave your bottles in a cooler part of the house (or even outside) before bringing them to the table.

Decanting

Most reds and even some whites are improved by decanting, since it gets oxygen into them and opens up their flavors. Decant reds several hours beforehand. Whites can be done just before serving. (See page 15 for how to decant.)

GLOSSARY OF WINE TERMS

Acid/acidity Acids occur naturally in wine and are crucial in giving it character and structure and in helping it to age.

Aroma The varietal smell of a wine.

Balance A wine's harmonious combination of acids, tannins, alcohol, fruit, and flavor.

Bereich (German) Term for a wine-producing district.

Bianco (Italian) White.

Blanc (French) White.

Blanc de blancs (French) A white wine made only from white grapes.

Blanc de noirs (French) A white wine made only from black (red) grapes.

Blanco (Spanish) White.

Blind tasting A tasting of wines at which the labels and shapes of the bottles are concealed.

Bodega (Spanish) Winery.

Body The weight and structure of a wine.

Botrytis cinerea A fungus that, when it shrivels and rots white grapes, concentrates their flavors and sugars, leading to dessert wines that are high in alcohol and richness of flavor. It is also known as noble rot, *pourriture* noble, and *edelfäule*.

Bouquet The complex scent of a wine that develops as it matures.

Cantina (Italian) Winery or cellar.

Carafe Simple decanter without a stopper.

Cave (French) Cellar.

Cellar book A useful way of noting what wines you have bought, from where and at what price, as well as recording when you consumed them and what they tasted like.

Cepa (Spanish) Term for vine variety.

Cépage (French) Term for vine variety.

Chai (French) Place for storing wine.

Chambrer From *chambre*, the French word for "room," the practice of allowing a wine gradually to reach room temperature before drinking.

Château (French) Term for a wine-growing property—chiefly used in Bordeaux.

Claret Term given to the red wines of Bordeaux.

Climat (French) Burgundian term for a particular vineyard.

Clos (French) Enclosed vineyard.

Colheita (Portuguese) Vintage. A term also used for single vintage ports.

Corkage charge Levied on customers in restaurants who bring in their own wine.

Corked Condition, indicated by a musty odor, where a wine is contaminated by a faulty cork.

Cosecha (Spanish) Vintage.

Côte (French)
Hillside of vineyards.

Cradle A wicker basket in which some restaurants present a bottle of undecanted red wine—usually a burgundy—to the table.

Crémant (French)
Semi-sparkling.

Cru (French)
Growth or vineyard.

Cru Classé (French)
Classed Growth, especially those 61 red wines of the Médoc (and one from the Graves) in Bordeaux that were graded into five categories determined by price (and therefore, in theory, quality) in 1855. Elsewhere in Bordeaux, similar classifications followed for the red wines of Graves in 1953 and for St Émilion in 1954 (revised in 1969 and 1985).

Cuvée (French)
Blended or specially selected wine.

Decanter Glass container with a stopper into which red wines and ports are decanted in order to allow them to breathe or to remove them from their sediment.

Demi-sec (French)
Semi-sweet.

Dolce (Italian) Sweet.

Domaine (French)
Property or estate.

Doux (French) Sweet.

Dulce (Spanish) Sweet.

Fermentation
The transformation of grape juice into wine, whereby yeasts—naturally present in grapes and occasionally added in cultured form—convert sugars into alcohol.

Fortified wine A wine—such as port, sherry, Madeira, or Vin Doux Naturel—to which alcohol has been added, either in order to stop it fermenting before all its sugars are turned into alcohol (thus maintaining its sweetness), or simply to strengthen it.

Frizzante (Italian)
Semi-sparkling.

Grand cru (French)
Term used for top-quality wines in Alsace, Bordeaux, Burgundy, and Champagne.

Halbtrocken (German)
Medium dry.

Horizontal tasting
A tasting of several different wines from the same vintage.

Jahrgang (German)
Vintage.

Keller (German) Cellar.

Landwein (German)
A level of quality wine just above simple table wine, equivalent to the French *vin de pays*.

Late harvest Very ripe grapes picked late when their sweetness is most concentrated.

Meritage Term first coined in 1988 for California wines blended from the classic red varieties of Bordeaux.

Méthode traditionelle
The method—involving a secondary fermentation in bottle—by which champagnes and top-quality sparkling wines are made.

glossary of wine terms 61

Moelleux (French) Sweet.

Mousse (French) The effervescence that froths in a glass of sparkling wine when it is poured, and which seems to wink at you.

Mousseux (French) Sparkling.

Négociant (French) Wine merchant, shipper, or grower who buys wine or grapes in bulk from several sources before vinifying and/or bottling the wine himself.

Non-vintage (NV) Term applied to any wine that is a blend of two or more different vintages, most notably champagne and port.

Nose The overall sense given off by a wine on being smelled. It is not just the wine's scent; the nose also conveys information about the wine's well-being.

Oak Much wine is aged in oak barrels, something which is typified by whiffs of vanilla or cedarwood.

Oxidized Term used to describe wine that has deteriorated owing to overlong exposure to air.

Perlant (French) Refers to a wine with the faintest of sparkles in it.

Perlwein (German) A type of low-grade semi-sparkling wine.

Pétillant (French) Slightly sparkling.

Phylloxera An aphid-like insect that attacks the roots of vines with disastrous results.

Punt The indentation at the bottom of a bottle, serving not only to catch any sediment but also to strengthen the bottle.

Quinta (Portuguese) Winegrowing estate.

Récolte (French) Crop or vintage.

Rosso (Italian) Red.

Rouge (French) Red.

Sec (French) Dry.

Secco (Italian) Dry.

Seco (Spanish/Portuguese) Dry.

Sediment The deposit that forms after a wine has spent a lengthy period in the bottle.

Sekt (German) Sparkling wine.

Sommelier Wine waiter.

Spittoon Receptacle into which wine is expectorated at a tasting.

Spritzer A refreshing drink made from white wine and soda or sparkling mineral water, usually served with ice.

Spumante (Italian) Sparkling.

Sur lie Term given to the process of ageing wines on their lees or sediment prior to bottling, which results in a greater depth of flavor.

Tafelwein (German) Table wine.

Tannin The austere acid—and necessary preservative—found in some red wines, usually young ones, which derives from grape skins and stalks combined with the oak barrels in which the wine has been aged.

Tastevin A small silver tasting dish, most commonly used in Burgundy, France.

Terroir (French) Literally, "soil" or "earth," but the term also encompasses climate, drainage, position, and anything that contributes to the mystery that makes one wine taste as it does while its immediate neighbors—grown and produced in the same way—taste different.

Tinto (Spanish/Portuguese) Red.

Trocken (German) Dry.

Ullage The amount of air in a bottle or barrel between the top of the wine and the bottom of the cork or bung.

Varietal A wine named after the grape (or its major constituent grape) from which it is made.

Variety The term for each distinctive breed of grape.

Vendange (French) Harvest or vintage.

Vendange tardive (French) Late harvest.

Vendemmia (Italian) Harvest or vintage.

Vendimia (Spanish) Harvest or vintage.

Vertical tasting A tasting of several wines from the same property that all come from different vintages.

Vigneron (French) Wine grower.

Vin de pays (French) Country wine of a level higher than table wine.

Vin de table (French) Table wine.

Vin Doux Naturel (VDN) (French) A fortified wine that has been sweetened and strengthened by the addition of alcohol, either before or after fermentation has taken place.

Vin ordinaire (French) Basic wine not subject to any regulations.

Vinification Wine making.

Vino da tavola (Italian) Table wine.

Vino de mesa (Spanish) Table wine.

Vintage Both the year of the actual grape harvest as well as the wine made from those grapes.

Viticulture Cultivation of grapes.

Weight The body and/or strength of a wine.

INDEX

A
Agiorgitiko 55
Albariño 38, 48, 55
Asian food 50–2

B
Barbera 16, 33
Beaujolais 16, 17, 27, 33, 38
Bordeaux 10, 18, 19, 33, 35
Buck's Fizz 26
Burgundy 19, 27, 35

C
Cabernet Franc 48, 51
Cabernet Sauvignon 10, 13, 19, 29, 33, 45
Carmenère 16, 29
Catarratto 48, 55
cava 25, 29
celebration wines 24–5, 28–9
Chablis 48, 50, 51
champagne 14, 15, 24–5, 26, 29, 35, 38, 40–1
Chardonnay 10, 13, 16, 19, 27, 28, 33, 45
Château Musar 55
Chenin Blanc 10, 13, 16, 28
Chianti 18, 51, 55
Côtes du Rhône 45
Crémant 25

D
decanting 15
dinner party choices 22
Docetto 52

F
Fitou 16, 55
Francs 17
French food and wines 46–7
Fumé Blanc 27, 28
fusion food 57

G
Gavi 48
Gewürztraminer 33, 50, 53
Grenache 16
Grüner Veltliner 19

I
Indian food 53
Italian food 48–9

L
Languedoc 54
Loire Cabernet 17

M
Madeira 15
Malbec 13, 16, 29
meat dishes 18–19, 26, 47
Merlot 13, 16, 19, 33, 45
Mexican food 54
Middle Eastern food 55
Monastrell 33
Montepulciano 48, 49
Muscadet 16

P
Pinot Blanc 10, 52, 53
Pinot Grigio 10, 33
Pinot Noir 10, 16, 17, 19, 27, 38
port 15, 40–1
Primitivo 48
prosecco 14, 25

R
restaurant food 44–5, 48–55
Rhône 53, 54
Ribera 18, 55
Riesling 13, 15, 16, 19, 27, 28, 33, 38
Rioja 10, 16, 18, 38
Rosé 17, 19, 38
Roussillon 49

S
sake 51
Sancerre 50, 51
Sauternes 15, 41
Sauvignon Blanc 10, 13, 16, 19, 33, 37, 38, 45
seafood dishes 46, 56
Sémillon 16, 19, 28
sherry 14, 15, 40, 57
Shiraz 10, 13, 19, 27, 29, 33, 45
Soave 16, 38
spirits 41
storing/serving 58
Syrah 13, 19, 27, 29

T
tannins, and heat 52
Tempranillo 33
Tokaj 53
Tokay Pinot Gris 33, 52, 53

V
Valpolicella 19, 33
vegetarian dishes 19
Verdelho 16
Viognier 16, 28

Z
Zinfandel 16, 27, 54

ACKNOWLEDGMENTS

Thanks to my parents for introducing me to wine at an early age, to my lovely wife, Bola, who has borne my vinous ramblings down the years with great good humor, and to my kids, Becky and Eddie, who make it all worthwhile.